GRE

Suzannah Evans lives in Sheffield and has published two poetry books, *Near Future* and *Space Baby* with Nine Arches Press. In 2019 she was a Gladstone's Library Writer in Residence, and in 2021 she received a Northern Writers' Award for poetry. Her work has been described as 'doom-pop poetry with an apocalyptic edge.'

First published in 2024 by Little Betty, an imprint of Bad Betty Press
Cobden Place, Cobden Chambers, Nottingham NG1 2ED

badbettypress.com

Copyright © Suzannah Evans 2024

Suzannah Evans has asserted her right to be identified as the author of this work in accordance with Section 77 of the Copyright, Designs and Patents Act of 1988.

PB ISBN: 978-1-913268-62-6
EPUB ISBN: 978-1-913268-63-3

A CIP record of this book is available from the British Library.

Book design by Amy Acre
Cover artwork by Nadiia Forkosh

Green

SUZANNAH EVANS

LITTLE BETTY

CONTENTS

There's Astronaut Shit on the Moon!	7
Green Finds Himself a Hundred Times	8
Earth was having nightmares	9
Origin Story	11
Nasty Nature	13
It is natural	14
Green on the Beach	15
Category Is: Extinct Species of the World	16
It's Spring	18
Green Frightens a Tree	19
Green's Rage	20
Green Helps the Neighbours Name Their Pets	22
Green and the Wasps	23
Against His Better Judgement	24
Equinoxes	25
Solstices	26
Self-Optimisation	27
Green and the Tincture Farm	28
Untranslatable Words	29
Green Goes to Plastic Island	31
Sizewell Tea	32
Acknowledgements	34

In that dream, the trees laugh at them.
Save us? What a human thing to do.
Even the laugh takes years.

Richard Powers, *The Overstory*

THERE'S ASTRONAUT SHIT ON THE MOON!

                                                        is the sort
of thought that keeps Green awake at night.
We've shored up so much stuff it makes him
grind his teeth. Above his head in low orbit
he feels the creaking halo of space junk,
a tinsel of broken satellites and rocket bits
that clank and roll and cinder back to Earth
where the problem's even worse, all of us
amassing armoires, Laliques, the enamelled
snuff-boxes of our ancestors.
In the pull-down dark of storage units
that stare across abyssal corridors,
paraphernalia waits; inhalers retire
with no puff left, old spectacles in drawers
tangle their stick-insect limbs.

GREEN FINDS HIMSELF A HUNDRED TIMES
A Tour of Foliate Heads

Sometimes Green wonders if he even is
a fertility symbol, if he isn't just some
stonemason's doodle, an exercise
in chisel-chipped mastery. In this chapel
his face looks down from every surface
into the vicar's cup of tea, a bible flipped
open on a lectern, the smiling iPhones
of tourists. Sometimes he looks
like he's screaming in pain. He's
a pebble-faced boy and a crumbling skull
biting down on yew roots. He's rebirth
and spring shoots, he waggles
his demon tongue. He's made
of leaves or lives among them,
deer-horned or bald, with fat
pinchable cheeks. In the gift shop
he's cufflinks, a bottle stopper
make-it-yourself cross-stitch pincushion.

EARTH WAS HAVING NIGHTMARES

and bad feelings.

She told the Moon, who got very worried.
Where would she be without the Earth?
The moon loved to gild her seas and forests
and neighbourhoods of tiny houses.

The Moon told Green, a natural
problem solver. Green drew up a list
of everyone, suggested an oath
for them to take

against harm, to put right the seasons,
to clear up the mess. Moon
swivelled sharp beams into their eyes
but there was no need for coercion:

>	the swallow, the gnat, the hare
>	the harrier, the oak, the mistletoe
>	the sundew, the hawkmoth, the great
>	white shark, the fern, the spores,
>	the soil and all its worms, the bees
>	the schist and slate, the brine
>	the rivers and deltas, the reeds
>	the snake, the salmon
>	(everyone consented)

The humans hold a summit every year.
They *can't possibly agree* without it.
They jet in from all their countries
while Green chews the skin off his fingers

>   re-reading their *out-of-office* replies.

ORIGIN STORY

My mother was an oak tree
        my dad a garage mechanic
My father was a field of wheat
        my mother the Prime Minister
My mother was an innkeeper
        and my father a lonely cactus
Mum was a hazel, Dad a tractor
        My father traded stocks and shares
My mother was a pampas grass
        My father was a student, a kid
who was yet to be anything
        my mother was a hawthorn hedge

It's a tale as old as time: human meets greenery,
talks a big game and something emerges. Me!

Now I live round the back
        of the household waste disposal site
listening to the jays shout
        about the acorns they love
My mother was a jay
        my father was a broken DVD player
consigned to the electricals bin

They call me *Green Man*
half green, half man
but I don't half feel estranged
from the city's rootless humans

The frogs in the pond turned leathery
        the bees froze to thistles in snow
that fell on May Day

So call me *Green*, sage, apple, teal,
        matcha, pine just the colour please,
whatever. No more Man for now.

NASTY NATURE

When the orca team up to hunt the seal
off its lozenge of ice, Green isn't rooting for anyone.
He watches the waves slap it over the side
its resignation as teeth close tight on tail.
Green is neutral, too, on the baby iguanas
on broken shell beach, running the gauntlet
of snakes with heads that sway like haunted seaweed.

The Komodo dragon and the camera crew
wait as the water buffalo grows weaker
until its legs give way and it slumps at last
in the mud of the watering hole. The dragons
are satisfied by this narrative arc —
and so are we, eating biscuits on the sofa.

IT IS NATURAL

for a prey animal

to be big-eyed and hypervigilant.

I was born like this. I wobbled

my way across a field before

I was two hours old. My family were scared

for me, they never let me walk home by myself

or watch *Bambi*. Sometime later

I came to a house in a city

where I tend houseplants

drink decaf to avoid the jitters

but I haven't forgotten which toadstools

are poisonous, or the deadliest of weathers.

most mornings I get up before anyone's awake

let my heart chase itself over garden fences.

GREEN ON THE BEACH

Wanted to be cool
sit on a deckchair
and look over his shades
at the waves
not saying anything
about how long your bucket
and spade might take
to biodegrade

but he couldn't do it —
he's litter-picking now
on the shoreline
putting starfish back
in safe places.
The hermit crabs choose
bottle tops; Green finds them
suitable shells

back and forth he goes
pulling a frayed
blue trawler rope
up and out of the sand
zigzagging wildly
interrupting picnics
never finding the end.

CATEGORY IS: EXTINCT SPECIES OF THE WORLD

Green drips with emerald scarabs,
wasp-waist cinched for the gods
in black feathers, lip-synchs

to the mysterious starling
a charcoal-smoke cabaret fantasy
before he rugs himself

like his printed sisters: quagga,
thylacine, Sumatran tiger
stripes strobing as they claw the air.

He runs his hands
down a velvet evening gown
as De Winton's golden mole

winks and guns us in the fashion camo
of the Yunnan lake newt. The moa
wears heels so high they're stilts

and the giant vampire bat
swirls in a silk black cloak,
a playful blood capsule

burst and streaking his chin.
The camera swoops to frame tears
that draw tracks

down silver fishskin cheeks.
What would you say to her today?
Steller's Sea Cow he answers

The world is a sadder place
without you; you were beautiful
and fierce, chunky yet funky

an ocean-going inspiration.
The final number is dedicated
to the endangered

with every chorus a gown peels back
and reveals a new shade of moth
or octopus. Green reaches out

towards us, eyes whiskered,
legs writhe in snakeskin.
The beat kicks, the lights

flash and stop. He slides a foot
in time. Amen. Death drop.

IT'S SPRING

and all of nature is doing it
sap is surging into extremities
the silver birches are full of it
daffodils bob their heads
snowdrops bend over
frogs creak like bedsprings
swans check themselves out
in the mirror of the canal

Green watches and they let him
they know he is single and singular
that the season makes him want
the soil takes his fingertips daintily
and kisses each one
rain licks the back of his neck

GREEN FRIGHTENS A TREE

The plum tree didn't mean
to get caught up in it all.
She was just resting

having produced
a heart-rending froth
of blossoms, standing

like a bride for the bees
and their hums
of spring gratitude

but Green needed
a listener; he'd been reading
the environment section

worries were piling up
jenga-ous and unstable.
When he told her

she shivered –
dropped her white veil
plums swelled bruisy and fast.

Green took a couple
of sweet ones
for the walk home.

GREEN'S RAGE

Is the only bright thing
in the city.
He burns a boron flame
against the white overcast,
the concrete buildings, black
bare trees, the black
slick of their fallen leaves.

Only the rain is see-through.
He stomps on.
Face of folded oak bark.
All the most aggressive weeds
are at his service.
He clenches and unclenches
root bole hands.

In the square outside
the city parliament
people are playing chess at tables
in wet black coats.
Feely tendrils of convolvulus
sprout to topple
knights and bishops.

A great thrust
of knotweed burgeons
in the inner chambers,
reaches the speaker's knees.

They will have to get the gardener
in again. Green kicks
the pavement.
It splits and ground elder
shoots out, bends
towards the drizzle.

GREEN HELPS THE NEIGHBOURS NAME THEIR PETS

They've only done the dog so far
and the hedgehog who snuffles the decking.
They've paid no attention to the nest
of slugs under the sink, the spider
in the corner of the window
eating anonymous fruit flies
who admittedly are numerous
and almost too ephemeral to christen.
Green names each
in its own language and outside
a rat runs up the wall, which is a laugh
because the big secret is
rats think humans are disgusting
with their grandiose buildings
and systems of oppression.
If you want anarchy, ask a rat,
clinging to a breadcake in the canal,
eating his raft as it goes down.
Anyway the couple are sleeping,
about to be woken by the dog
shouting its own true dog name
out of the letterbox
in case someone is passing.

GREEN AND THE WASPS

No-one likes you, Green says,
but I get it. If you didn't
no-one else would. If not you
then who, if not now, when?
Etcetera. I appreciate you,
wasps, especially
at the end of summer
when you've lost your holiday jobs
and you're staring down
one last season before the clamp
of winter. Why wouldn't you
climb inside a fermenting plum
and get wasted? The wasps
sting Green's skin into a cactus
of goosebumps. Fuck you
says Green, I still love you.

AGAINST HIS BETTER JUDGEMENT

Green loves all humans, too –
the idiot birdwatchers that sit
in the rain with telescopes
on a tip-off for a great grey shrike.
The people in the park don't notice
Green, how he's totally
in love with their new boots
and their rosy outdoor cheeks
or how the squirrels are super grateful
for the peanuts, the birds overhead
glad to see them, glad they've worn their hats
after all, the weather's turning.

EQUINOXES

spring

Daffodils feel hopeful and, when the tulips come up too, the soil is ecstatic but don't leave your waterproof at home just yet. Birds carry sticks across the park to their constructions and allotments tingle with potential as they plant up seeds in trays and repair greenhouse panes that the storms have cracked. Magpies and crows are always in pairs now, which skews everyone's thinking towards optimism. Thatnewjobisyours! Yourcrushfindsyoudelightful! Joy!

autumn

Did anyone make the most of the summer? Nihilist wasps drift around the parks, looking for a jam sandwich or a fight. Reservoirs hold some of summer's warmth and swimmers stride armpit-deep, wearing neoprene boots and gloves. The water company sends security guards with alsatians who long to jump in, but don't. The leaves are thinking of all the colours they'll be, wondering where they'll land.

SOLSTICES

winter

Hail Smiling Morn! Carol singers stand in layers of wool and a weak lemon standstill light shows itself through a gap in the drizzle. A crowd gathers and everyone stands at a reasonable distance from each other. Many have been indoors in enforced hibernation, each day momentary in the long blue-black stretch of the season. Some wipe their eyes, stamp their feet to get the feeling back. *At whose bright presence / the darkness flies away!*

summer

Rising light hits one lake then another and deep gold crawls up the valley. On top of the mountain the walkers wait, sipping coffee from a thermos flask, gore-tex zipped to their necks. They imagine the towns and cities they come from, rows of houses, only insomniacs awake there. The raven family on the crag shuffle their wings, close their eyes for a bit longer.

SELF-OPTIMISATION

the early bumblebee aspires
to five more flowers than yesterday

the hedgehog knows outside the hedge
is where the magic happens

the ivy is in training
to overtake the garage by spring

the moon puffs up perigee pink
and full of confidence

the adder strips its skin
in preparation for a sensational new life

the wood ants dream big
with this season's pine needle heap

the hare hits a personal best
high fives himself so hard he flips

over the curled fox, who isn't playing –
asleep in the pizza boxes and autumn leaves

GREEN AND THE TINCTURE FARM

Green lies back in the lavender field
at the edge of the organic farm
the sun and bees are soporific
he wiggles his jaw loose
the yellowhammers are singing
the lavender is grown
for a self-care tincture
called *fatigue fix*. The women
in the city love it, they drip drops
under their tongues at their desks —
they spent £22.50 on it
it must be good for them —
resume their spreadsheets
try not to think about sinking
to the lilac-grey office carpet
to feel its ridges on their cheeks
inhale sweet restful summer
a birded hedge
a felted skyful of hum.

UNTRANSLATABLE WORDS

Green knows every language
all the human ones
and the semiotics of wind

when it swears and blows
the dustbins over, or rain
as it scolds the tarmac

looking for somewhere to run.
Green goes with pad and pen
to the forest, takes dictation

from the bubbles that pop
on the surface of the pond
each one loaded with words

from the mud below.
He listens to the battle cry
of brave newts taking to land

the psithurism of lime trees
the garlic breathiness of ramsons
the sticky burp of a mistle thrush

in berry season. He takes the words
to the humans who love them,
get him to write a listicle:

Twenty-Five Super Fresh
Untranslatable Words from Nature
That You've Never Heard Before!

His byline photo is an assemblage
of three knots in some wood.
It's the most-read story that day

until the top spot slides
to *Forty Times Llamas*
Were Devastatingly Majestic!
and he has to admit they were.

GREEN GOES TO PLASTIC ISLAND

and straddles a crackly nest of
bobbing evian bottles, puts his
feet up on a TO LET sign
he knows the heap of trash
within this widening gyre
outweighs
our collective biomass

he watches the sun go down
psychedelic with ice-cube trays
and flame-pink cellophane

the island is alive
with crepuscular beings –

tyre-rubber turtles
tootling between mounds
of blister packs

tired sharks whose fins
flap like vinyl flooring

snakes and ropes in courtship
weaving frenzied kelp

shoals of jellyfish carrier bags

floating angels everywhere   watch the stings

SIZEWELL TEA

Flora and Fauna Flourish
says a sign, which has rusted
in the spray. Green doesn't
doubt it — sea plants tuft
the shingle, even on a day
like this one, when the wind bites.

There's a bird reserve
nearby – Caspian Gulls
drift in, displaced
from the banks of the *Dnipro*.
All down the shore sea-anglers
are folding up their little tents
shoving wet fins into iceboxes.

Seals surface doggy heads to stare
ocean-eyed at buildings A and B
who flank each other, the old one
a textured concrete cube
the new one the colour of sky
its white dome shining
like a pickled egg.

Green has been told you can't feel
radiation, but it's there
hovering in little stink-lines
above the grass. Someone
has buried their dog on the bank

under a polished wooden cross.
In a century's time
its bones will be gone

but Green will hear
the land's invisible clicks
for another ten thousand years.
Will the tea-shop's name
still be ironic then? Will
the fish be taking the bait?

ACKNOWLEDGEMENTS

I am grateful to the Sheffield Writer Development Fund for a grant that helped me fund the time and research to write these poems.

Thanks to those who have read and offered feedback on these poems; Genevieve Carver, Tom Sastry, Roy Marshall and Harry Man, and to the members of Zellig Poetry Group.

The poem 'Solstices' reproduces some of the lyrics to Hail Smiling Morn, a carol or glee by Reginald Spofforth, traditionally sung at Christmas and Easter in the north of England, especially Sheffield. The poem 'Category is: Extinct Species of the World' owes a similar debt to RuPaul's Drag Race.

Thanks to Jake and Amy at Bad Betty Press, and my editor Anja Konig for all your work on this pamphlet.

Milton Keynes UK
Ingram Content Group UK Ltd.
UKHW012137230524
443117UK00004B/128